P9-CFN-653

FLIGHTS of IMAGINATION

AN INTRODUCTION TO AERODYNAMICS

BY WAYNE HOSKING

NATIONAL SCIENCE TEACHERS ASSOCIATION

Acknowledgements

Special thanks to the following people whose expertise and invaluable contributions have made this book possible.

Herb Birnn
Margo Brown
Margaret Greger
Robert Jones
Richard Reif
Charlotte Ward

Copyright© 1987 by the National Science Teachers Association
1742 Connecticut Avenue, NW, Washington, DC 20009

ISBN Number: 0–87355–067–6

This volume has been produced by Special Publications
National Science Teachers Association
1742 Connecticut Avenue, NW
Washington, DC 20009
Edited by Shirley Watt, managing editor
        Alice Green, assistant editor
        Mary Liepold, editorial assistant
Illustration by Loel Barr
Design by Barbara Sahli

CONTENTS

HISTORY

No one knows what the first kites looked like. Historians generally accept, however, that the first kites were flown in China about 3000 years ago. Throughout Asia, kites were used for recreational, religious, and military purposes. There are records from the Han Dynasty (200 B.C.–200 A.D.) of the Chinese military attaching bamboo pipes to kites flown over enemy troops. As the wind passed through the pipes, the whistling sound caused troops to panic and flee. Kites reached Europe around the fourteenth century via the trade routes.

Over the centuries, kites have been used in a range of scientific endeavors. The first reference to a kite being used for measurement comes from China. In 196 B.C. General Han Hsin used kites to measure the distance from his troops to the enemy's stronghold. In 1749 Alexander Wilson measured the variations in temperature at different altitudes using a train of kites. In 1752 Benjamin Franklin used a kite to conduct electricity from lightning down a wet line to a key, showing that natural lightning is electricity. (Historians still wonder how he survived!) In 1847, a kite helped pull a cable across the Niagara river between the United States and Canada to form a link in that river's first suspension bridge. William Eddy's kites, in the 1890s, were used to raise meteorological payloads and contributed greatly to the science of weather forecasting.

It has been in the development of aeronautics, however, that kites have contributed the most to science. The oldest basic type of kite is the flat kite. Though the flat kite has taken many shapes, and has been made enormous in size, it was not improved on until Eddy added the bow to the kite, creating a positive dehedral angle to the wind and increasing kite stability. From this start, kite design progressed to the box kite, developed in Australia by William Hargrave. Box kites provide much more lift than flat designs, and Hargrave experimented in 1893 with trains of box kites which generated enough lift to raise a man off the ground. The Wright brothers experimented with box kites to test wing performance for the design of the first airplane. And Alexander Graham Bell developed enormous tetrahedral box kites for carrying people. Bell's kites only carried people above the lake near his experimental station in Nova Scotia—the development of Wright's airplane in 1903 proved to be a better design. Bell still used the tetrahedral kite design to create the first hydrofoil boats, which aided the Allied effort at the end of the First World War.

Though kites have been used as tools for centuries, how kites really work has still not been thoroughly investigated. We know that the forces of lift, drag, and gravity combine to keep a kite in the air. But as you proceed through the projects in this book, you will discover that subtle changes in design can make a big difference in the way these forces interact.

PARTS OF ALL BASIC KITES

The FLYING LINE holds the kite captive to the wind.

The BRIDLE connects the kite to the flying line, at the TOW POINT, and sets the angle of the kite to the wind.

The SPINE is the backbone of a kite.

The STRUTS are the side and cross stricks that keep the kite open.

The FRAME is the combination of the spine and strut(s) that supports the kite sail.

Delta Kite Parts

The KEEL is the backbone of a delta-shaped kite.

The SPREADER is a strut perpendicular to the keel.

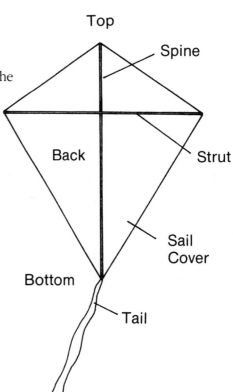

DIAMOND KITE

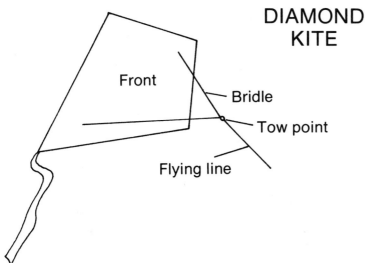

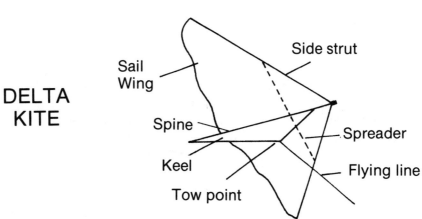

DELTA KITE

7

FLIGHT DYNAMICS

Flat kites include diamond and hexagonal kites. They fly at a low angle and require a tail for stability. The tail is to provide yaw stability by adding drag, and not to add weight to the back of the kite.

Flat plane

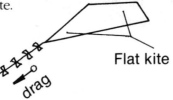

Flat kite

drag

Eddy and bowed kites have their cross struts set in a positive dihedral for roll stability. They do not require a tail to fly.

Dihedral angle

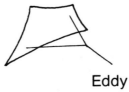

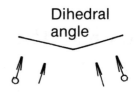

Eddy

Delta kites also form a positive dihedral for roll stability. The keel adds yaw stability. They are high angle fliers.

Delta

Sled kites form a negative dihedral. The side keels add yaw stability. Shaped vents or tails are sometimes used for extra stability. The bridle should not be shortened on this kite, or the kite will collapse.

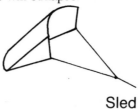

Sled

Fighter kites on a slack line have limited stability. The kite has a flexible cross strut. When the flight line is taut, it bends, forming a positive dihedral. When the line is slack, the kite flattens. These kites are designed to be maneuvered and are usually flown tailless.

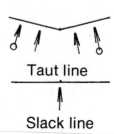

Taut line

Slack line

Fighter

Stunter kites form a positive dihedral. The bridle and double line arrangement allow the flier to control these kites. When one line is pulled, the leading edge dips and the kite travels in that direction. The dihedral provides roll stability during the maneuvers. The spine should have a slight curve to provide pitch stability and stop the kite from gliding.

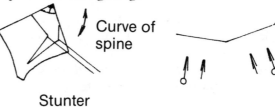

Curve of spine

Stunter

The cells of a box kite form positive and negative dihedrals for roll stability. The distance between the cells is important for pitch stability.

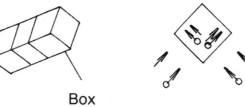

Box

Double box kites fly flat. The vertical sides act as yaw stabilizers.

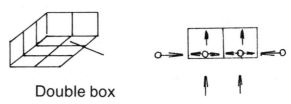

Double box

Winged box kites have the same stability as box kites, but with extra lift. The cells of a triangular box form a positive dihedral.

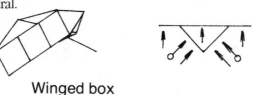

Winged box

Tetrahedrals fly like box kites but with two lifting surfaces instead of four. They have a positive dihedral and are usually flown tailless. They require a steady wind and are difficult for a novice to fly. Tetrahedral kites can have hundreds of cells.

Tetrahedral

Aerodynamics

A kite is an object that is flown in the wind on a line. It is usually made of a light frame with a thin covering. Technically a kite is a tethered aircraft set in a stall. A hang glider becomes a kite when a tether or line is attached.

Daniel Bernoulli (1700-82), a Swiss mathematician, formulated a principle that explains some of the aerodynamic principles of AIRFOIL (airplane wing). This principle states that when air moves over a surface, the greater the velocity, the lower the pressure on that surface.

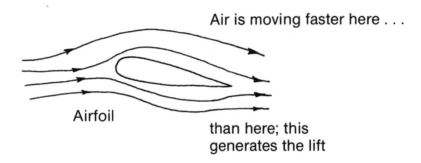

Air is moving faster here . . .

Airfoil

than here; this generates the lift

Bernoulli's principle, which accounts for lift of kites as well as other airfoils, is really just a statement of energy conservation in the case of streamline flow. At every point in the air stream, the total energy must be the same. The three kinds of energy to be accounted for are "volume energy," kinetic energy, and potential energy. If the pressure of a volume of gas is reduced, then the kinetic energy must increase to compensate for the decrease in volume energy. This occurs by the increase in velocity of the gas.

A flat surface at an angle—a kite—acts substantially like an airfoil as far as Bernoulli's principle is concerned.

Pressure and force are not the same thing. Pressure is force per unit area. The difference in air pressure on the two surfaces of the airfoil results in a net upward force called lift. All forces may be assumed to act at a point called the center of mass of the kite or aircraft.

Low pressure

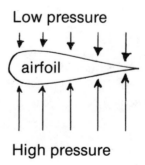

airfoil

High pressure

Kite Aerodynamics

LIFT results from differential air pressure on the kite's surfaces.

DRAG is created by air friction along the kite's surface. Further drag can be created with the addition of a kite tail, because every additional surface increases the frictional force.

GRAVITY is the downward pull created by the weight of the kite, flying line, and tail. For a kite to fly, there has to be enough lift to overcome gravity.

The three main forces (LIFT, DRAG, and GRAVITY) are at work on a kite in flight acting on a single point of balance at the center of mass.

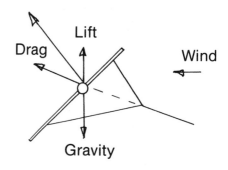

The position at which a line attaches to a kite (the tow point) has an effect on the STABILITY of the kite in both lateral (pitch) and vertical (yaw) axes. Stability can come from the drag of a tail or a built-in feature such as a DIHEDRAL. When a kite with a dihedral dips to one side, greater force is exerted on that side, returning the kite to equilibrium. Such a kite has roll stability.

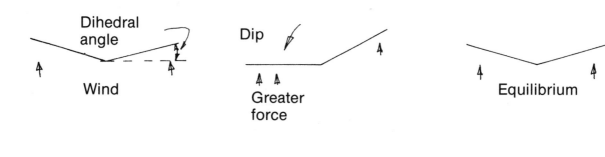

Dihedral

PROJECT 1: *WHAT WILL FLY?*

Question

Does the shape of an object affect how it flies?

Materials

A round balloon
A Gayla inflatable kite, or "puffer kite"
Flying line

Procedure

Inflate the balloon and inflatable kite and attach flying lines. Compare the reactions of the balloon and kite in the wind.

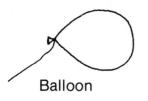

Balloon

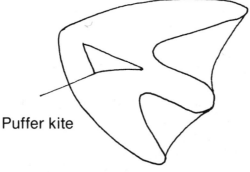

Puffer kite

Think About It

- How do the balloon and the kite react to the wind?
- Is it possible to make a balloon fly? Why not?
- Can a brick be made to fly? How?
- Can you make a list of things that will fly?
- Can you define flight?
- What shapes can you imagine flying? There is plenty of inventing yet to be done in the field of aerodynamics.

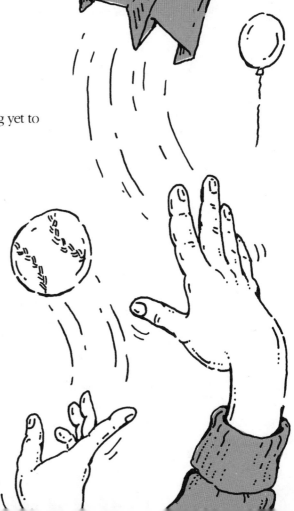

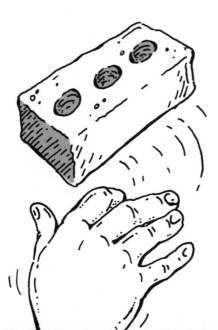

PROJECT 2: *KITES AND GLIDERS*

Kites and gliders are both *aircraft* which require an external energy
source to fly, but:
—A kite relies on wind speed to overcome gravity, while a glider uses
gravity to give it gliding air speed.
—A stable flying kite is set in a stall, while a stalled glider will fall out of
its glide path.

Question

Do gliders and kites fly the same way?

Procedure

Build a simple paper kite and a glider to compare their flight characteristics. Choose one kite and one glider.

Materials

Dingbat Kite

Sail	8 1/2" x 11" (22 cm x 28 cm) bond paper
Flying line	Sewing thread
Tail line	12" sewing thread (30.5 cm)
Bridle	18" sewing thread (46 cm)
Tail	2 — 1 1/4" x 6' 2 mil plastic strips (3 cm x 2 m)

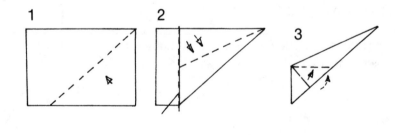

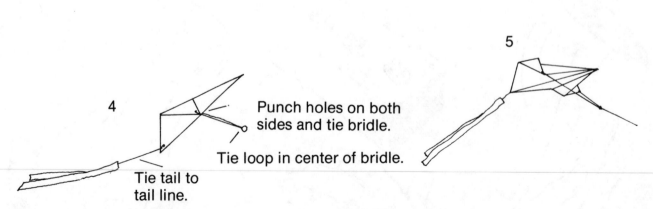

Punch holes on both
sides and tie bridle.

Tie loop in center of bridle.

Tie tail to
tail line.

Dart glider

8 1/2" x 11" (22 cm x 28 cm) bond paper

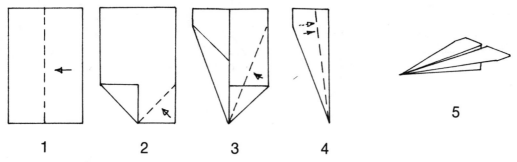

1   **2**   **3**   **4**   **5**

Fold and unfold

Cub kite

Sail	8 1/2" x 11" (22 cm x 28 cm) bond paper
Cross strut	8" x 1/8" (20 cm x 3 mm) dowel or drinking straw
Flying line	Sewing thread
Tail	3 — 1 1/4" x 6' (3 cm x 2 m) 2 mil plastic strips
Tape	5 pieces

1   **2**   1 1/2"

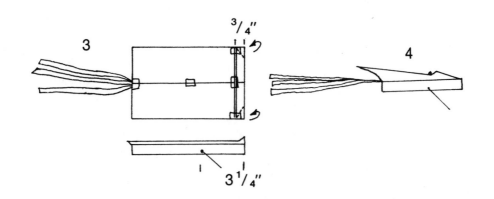

3   **4**   3/4"   3 1/4"

Glider

8 1/2" x 11" (22 cm x 28 cm) bond paper

Fold and unfold

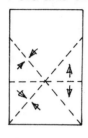 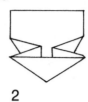 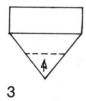 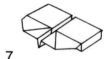

1

2

3

4 Fold in half.

5 Fold "wings" out, leaving a $^1/_2$" "keel."

6 Fold ends of wings up $^1/_2$" in the same direction as the keel.

7

Think About It

- Can you explain, in your own words, the differences between a kite and a glider?
- Can you fly a kite as a glider?
- Is it possible to tether a glider and fly it as a kite?

PROJECT 3: *LIFT*

Bernoulli's Principle states that fast-moving air exerts less pressure than slow-moving or still air. The curved surface of an *airfoil,* or airplane wing, causes wind to accelerate over the top of the wing and to decelerate over the bottom. The result, less pressure on the top of the *airfoil,* creates *lift.*

Question 1

Can you demonstrate Bernoulli's principle?

Materials

Strip of light card (e.g. a large file card, cut in half lengthwise.)
Drinking straw
Pencil

Procedure

Bend the card to the slightly curved shape of an airfoil. Fold the end over so that it will hang on the pencil. Using the straw, blow a steady stream of air over the top of card.

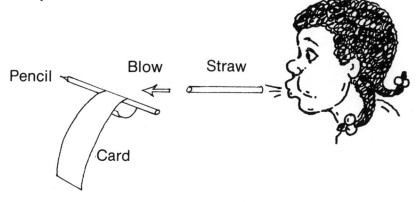

Think About It

- What do you see the card doing?
- Can you explain the results in your own words?
- What happens if you do the same thing with a single ply of tissue from a facial tissue?

Question 2

What effect does wind have on a flat plate?

Materials

Sheet of 8 1/2" x 11" (22 cm x 28 cm) paper

Procedure

Hold the sheet of paper by the top corners, and blow into the paper.

Think About It

- How does the paper react to the "wind"?
- Can you explain the results in your own words?
- What other experiments can you conduct that will compare the function of an airfoil to that of a flat plate?

PROJECT 4: *AIRFLOW*

A flat plate is an airfoil, but not as efficient an airfoil as one with curved surfaces. A flat plate does not provide as good a core for the air to flow smoothly around, so the pressure differences which create lift are not as strong as on an airfoil with curved surfaces.

An airfoil stalls when the air flow around it becomes turbulent. A stall is the condition in which an airfoil loses the speed necessary for control. The air flow around a flat plate becomes turbulent at a smaller angle of attack than for a curved airfoil. (Turbulent air has forces acting in random directions, providing no steady lift.) Any airfoil will stall: the angle of attack at which this happens will vary from one airfoil to another, depending on shape.

Question

How does wind affect a curved airfoil differently than it does a flat plate?

Materials

2 – 5" x 2" x 1/4" (13 cm x 5 cm x 6 mm) balsa wood or foam blocks
16 – 1/4" x 3" (6 mm x 8 cm) strips of tissue paper
Glue, pencil, and sandpaper
Wind source (e.g. hairdryer)

Procedure

1. Sandpaper or cut one of the blocks to resemble a cross section of an airplane wing. Leave the other as it is.

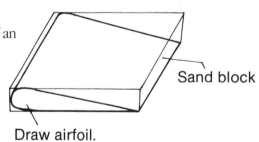

Sand block

Draw airfoil.

2. Glue 8 strips of tissue paper along the leading edges of both the wing and the flat plate.

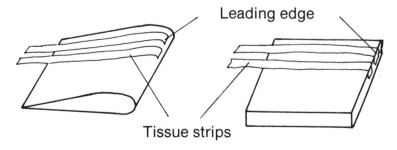

Leading edge

Tissue strips

3. Lawrence Hargrave used a curved airfoil and a flat plate in his 1893 experiments in search of a flying machine. You may wish to support your two airfoil shapes on stiff wire or straightened paperclips in an approximation of the method Hargrave used. Press the points of the wires into the airfoils just until they are supported. Hold the other ends of the wires level by inserting them between books or on a more permanent support of your own design.

Hargrave's Experiment

4. While blowing air on the leading edge, raise the airfoil, changing its angle of attack. Observe the reaction of the tissue strips to the air flow at a small (low) and a large (high) angle of attack.

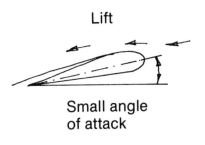

Lift

Small angle
of attack

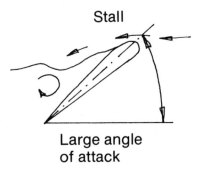

Stall

Large angle
of attack

5. Repeat the above process using the flat plate.

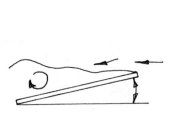

Small angle
of attack

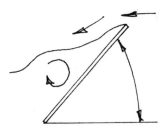

Large angle
of attack

Think About It

- Can you describe, in your own words, the result from the above experiment?
- Do all airplane wings have airfoil cross sections of the same shape?
- Compare the cross section of the Wright Brothers Flier to that of a modern jet.

PROJECT 5: *ANGLE OF ATTACK*

A kite is set at the correct *angle of attack* to the wind by adjusting the *tow point*. The tow point is the point where the flying line attaches to the bridle or directly to the kite. Increasing the angle of attack will increase *lift* until a negative force, called *drag*, takes over and makes the wind ineffective. At this point the kite no longer provides lift and is said to be *stalled*.

As a kite rises, in an arc, the angle of attack changes. The final flying angle will depend on the tow point and the efficiency of the kite's lifting surface.

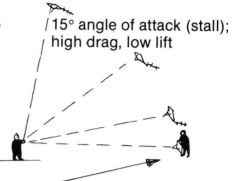

15° angle of attack (stall); high drag, low lift

70° angle of attack (launch); high lift, low drag

Question

What effect does the angle of attack have on how a kite flies?

Materials

One or more kites with a bridle or keel
Protractor (to read angles)

Procedure

To test the angle of attack, hold a kite at the tow point over a flat surface, and measure the static angle. The static angle is the angle formed between the kite and a horizontal surface when the end of the kite touches that surface. Change the static angle each time the kite is flown and note the effects this has on the kite's flight. Also note the static angle of a number of other kites and at what angle they fly.

Think About It

- What effect has the angle of attack have on how a kite flies?
- Do all kites fly at the same angle?
- Can you design a self-adjusting bridle using a small spring or rubber band?

NOTE: A tow point set at a *high* angle allows most wind to pass by the kite, but can also cause spinning and diving. If the tow point is set at too high an angle the kite will not lift, but just flutter.

Flight angle

Launching angle

Bridle

Static angle

High

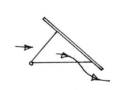

Creates lift

A tow point set at a *low* angle allows most of the wind to be caught by the kite's cover. However, if the tow point is too low the kite will refuse to fly. Another effect of too low a tow point is yawing, or pulling to one side.

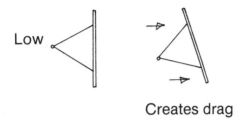

Creates drag

To test the tow point on a small kite have the kite make a figure eight. To do this, hold the kite by the tow point, out away from your body, and move the kite in circular motions. If the kite flows along, the tow point is correct. The tow point should be just above a point at which the kite refuses to follow along.

PROJECT 6: *MEASUREMENTS*

Measurements are very important in designing and making a kite or any other aircraft. A change in any of the kite's dimensions might dramatically change how, or even if, a kite will fly. In this project you'll investigate the properties of a sled kite.

Question

What effect do different measurements have on a kite?

Materials

1 – 24" x 30" (61 cm x 76 cm) heavy duty trash bag
2 – 3/16" x 24" (4 mm x 61 cm) dowels (struts)
72" (183 cm) flying line (bridle should be 3 times kite length)
Tape – filament, packaging, or electricians'

NOTE: This sled can be drawn using 20 cm as a unit, and dividing the bag into 4 units wide x 3 units high.

NOTE: To cut out your kite, hold the plastic taut. Barely open the scissor blades. Begin a cut and then pull against the scissors.

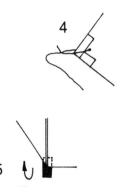

Procedure

Step 1

1. Draw half of kite on unopened bag and cut out sail. Open kite.
2. Tape dowel to kite.
3. Tie bridle to bridle points.
4. Find center of bridle by placing keels together. Tie a loop in the midpoint of the bridle. Attach the flying line to this point.

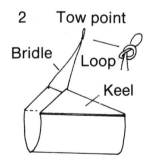

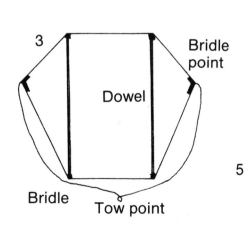

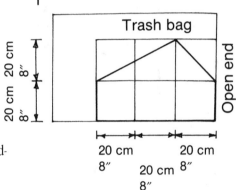

Tape under sail and fold onto dowel.

5. Slit a popsicle stick at both ends and place it between the bridle lines to stop them from twisting.
6. Add a tail if weather conditions require.

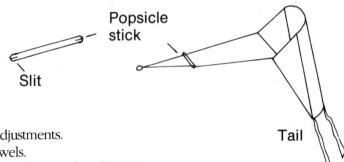

Step 2

Try flying the kite, making the following adjustments.
1. Use 1/8" (3 mm) and 1/4" (6 mm) dowels.
2. Use a short (e.g. 24" [60 cm]), then a long (e.g. 12' [3 m]) bridle.
3. Relocate the bridle points to a higher and then to a lower position.
4. Try relocating the tow point so that it is off-center.

Think About It

• What effect do different dimensions have on the way a kite flies?
• Can you make a sled kite using your own measurement? Hint: Start with 4 units wide and 3 units high. This will need a bridle length of 9 units. For a small kite use 1/8" (3 mm) dowels for struts. For a large kite use 1/4" (6 mm) dowels. Move the bridle points until the kite flies.

24

PROJECT 7: *DIHEDRAL*

One of the methods used in kite construction to add stability is to bow the wings or set them at an angle. This angle is known as a *dihedral*. When a kite with a dihedral dips to one side, greater force is exerted upward on that side, returning the kite to equilibrium.

Question

Does a dihedral stabilize a kite?

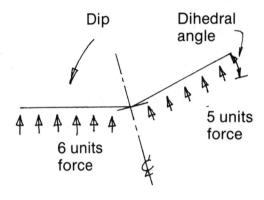

Materials

1 – 24" x 30" (61 cm x 76 cm) trash bag (sail)
1 – 3/16" x 24" (4 mm x 61 cm) dowel (spine)
2 – 3/16" x 12" (4 mm x 30.5 cm) dowel (struts)
1 – 3/16" x 12" (4 mm x 30.5 cm) dowel (spreader)
3 – 3/16" ID x 2" (4 mm x 5 cm) plastic tubing

Construction

1. Draw and cut out kite sail. (One plastic bag will make two sails.)
2. Center a hole all the way through one length of tubing, using a drill, leather punch, or scissors. (Keep it small—the tubing will stretch.) Insert spine and both struts.
3. Fold each of the remaining tubes in half and nick a small hole in the center, only cutting the outside. Slide one onto each strut, pushing dowel in through the end and out through the hole. The two loose sleeves will hold the spreader.
4. Align frame symmetrically and center spreader. Tape frame to sail.
5. Attach flying line 1 inch (2.5 cm) below the point where the struts and spine join.

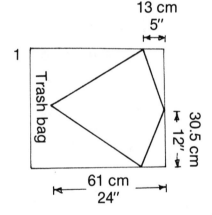

2 Fold each tube in half. Nick small corner.

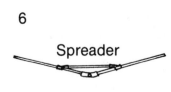

Side tubing (2)

3 Slide dowel in through end and out through hole.

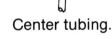

4 Make one hole all the way through the tubing.

Center tubing.

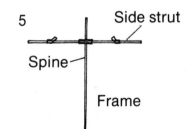

6

Spreader

7

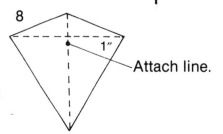

Tape under sail and fold onto dowel.

8

Attach line.

25

Procedure

1. Carefully try to fly the kite flat. Do not use a tail.
2. Keep moving tubing along cross struts (moving the same distance on both sides each time) and testing until the kite flies. Measure the dihedral angle.
3. Using a proctractor, measure the dihedral angle on a number of other kites.

Think About It

- Are all kites set at the same dihedral angle?
- Can you explain, in your own words, how a dihedral works?
- Draw the kite, at a reduced scale, on graph paper with one side dipping to illustrate the difference in lifting surfaces of each side.
- What other aircraft use dihedral angles? Notice that not all dihedral angles are positive.

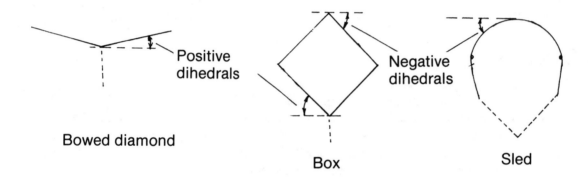

Positive dihedrals

Bowed diamond

Negative dihedrals

Box

Sled

PROJECT 8: *TAILS*

The main function of a tail is to add drag, not weight. The drag produced by the tail keeps the kite properly positioned into the wind, so that if a kite pulls to one side, the drag pulls in the opposite direction to reduce the lateral movement of the kite.

Kite tails do add weight, however, and shift the center of mass on the kite. A good tail should not throw the kite off balance so much that it cannot be corrected by adjusting the tow point.

Not all kites require a tail, and not all tails act upon a kite in the same way. In this project you will investigate types of kite tails and how they work.

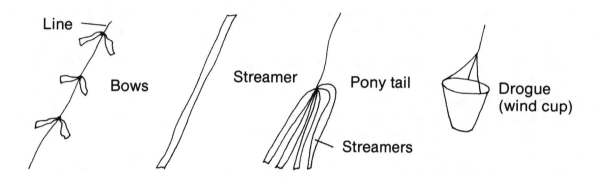

Question

What effect does a tail have on how a kite flies?

Materials

Kites
Trash bags to cut into 2" (5 cm) wide strips

Procedure

1. Make streamers—Roll trash bag and cut 2" (5 cm) wide strips.

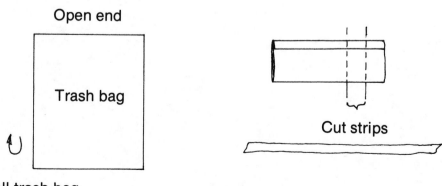

Roll trash bag

2. Make a drogue. A drogue is also called a wind cup or wind sock. The advantage of a drogue over other types of tails comes from its cup shape, which offers varying degrees of wind resistance in direct proportion to the wind velocity. In other words: the stronger the wind, the greater the resistance. To make a drogue, draw the pattern below on a plastic bag and cut it out. Secure the drogue to the kite from three points on the drogue, as shown. You may want to reinforce the leading edge of the drogue with packing tape.
3. Set up a test to study the effects of tails on a kite.

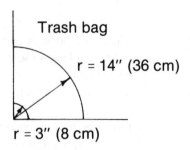

Trash bag

r = 14″ (36 cm)

r = 3″ (8 cm)

Tape

Think About It

- When is a tail necessary?
- Does a single long streamer tail—2" x 36 ft (5 cm x 11 m)—have the same effect as two short—2" x 18 ft (5 cm x 5.5 m)—or three shorter—2" x 12 ft (5 cm x 3.7 m)—tails with the same surface area?
- What effect does a tail have upon the amount of wind needed to lift a kite?
- What effect does a tail have upon the maximum wind a kite will take?
- How does a tail affect the angle of elevation? See Project 11.
- What tail gives the highest angle of elevation?
- What is the best tail for a kite with no dihedral (e.g. a flat kite)?
- What is the best tail for a kite with a positive dihedral (e.g. a bowed diamond or box kite)?
- What is the best tail for a kite with a negative dihedral (e.g. a sled kite)?
- What is the effect of changing the center of mass? Tape a small fishing weight (a sinker) to the bottom of a kite and note the effect. Tape a small fishing weight to the top of the kite and note the effect.
- What materials could be used as kite tails? What characteristics should a good kite tail have?

PROJECT 9: *MATERIALS*

The main factors to consider when choosing kite materials are aesthetics, the strength to weight ratio, ease of construction, cost, and wind resistance.

Different cover materials can be used on any given design as long as they are of similar weight and wind resistance. A heavier sail than is recommended might cause the kite to fly poorly or not at all. A kite made with only strength taken into consideration most likely will be too heavy to fly.

Question

What effect do different materials have on a kite's durability, construction time, and flight performance?

Materials

Newspaper (sail)
2 – 3/16" x 24" (4mm x 61 cm) dowels (spine and strut)
Line to frame the kite with (30# test)
30–40 feet (9–12 m) x 2" (5 cm) plastic tail
Glue

Procedure

1. Assemble frame and attach the framing line.
2. Lay the frame on the newspaper and trace the kite outline. Allow 1/2"
   to 1" (1 to 3 cm) flap for gluing.
3. Cut out and glue sail to frame.
4. Attach the flying line 1" (2.5 cm) below the point where the spine and
   strut cross.
5. Attach a tail to bottom of kite.

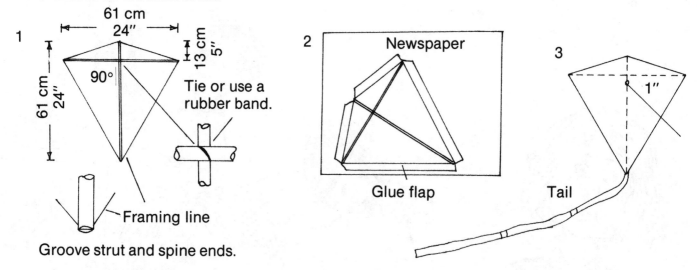

Think About It

- How do the flight characteristics of this newspaper kite compare with
  the plastic kite from Project 7?
- How is a kite affected by the materials used in its construction? Test and
  evaluate the following kitemaking materials.

Sail	Struts
Silk	Dowel
Sailcloth (ripstop)	Bamboo
Polyethylene	Cane
Cotton	Fiberglass
Nylon	Fiberglass tubing
Tyvek®	Aluminum tubing
Paper	Plastic tubing (aquarium hose)
Mylar®	Wire

PROJECT 10: *BOX KITES*

The *box kite* was a byproduct of Lawrence Hargrave's 1893 experiments with kite forms, in his quest for a flying machine. It is a three-dimensional unit whose sides can be squares, rectangles, or triangles.

Question

Will different configurations of the same box kite have any effect on how each kite flies?

Materials

For each kite:
1 – 24" x 30" (61" x 76 cm) trash bag
4 – 3/16" x 24" (4 mm x 61 cm) dowels (struts)
4 – 3/16" x 16 1/2" (4 mm x 42 cm) dowel (cross struts)
8 – 3/16" ID x 1 1/2" (4 mm x 4 cm) plastic tubing

Construction

KITE 1

1. Cut across the bag to form a single cell 24" (61 cm) long. Do not slit sides open.
2. Punch or cut a hole in tubing and slide onto struts.
3. Fold cell and mark each corner at each end.
4. Tape side struts inside the cell at the marked corners.
5. Fit cross struts into tubing on side struts.
6. Attach the flying line to the tow point 4" (10 cm) from the top.

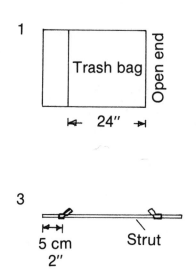

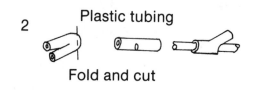

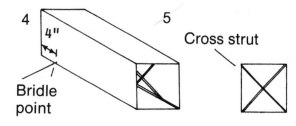

KITE 2
Same as KITE 1 but with two 8" (20 cm) cells.

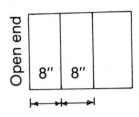
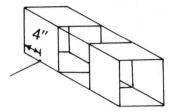

KITE 3
Same as KITE 1 but with two 4" (10 cm) cells.

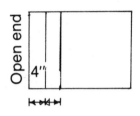
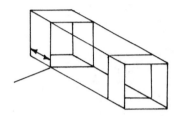

KITE 4
Same as KITE 1 but with vents.

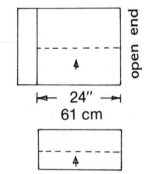
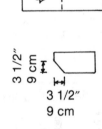
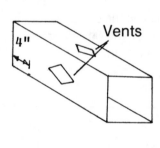

Procedure

Fly each box kite, noting any differences in flight.

Compare the flights of the box kites. Use a rating scale of 1–10 to rate each kite's performance. Which flew best?

Think About It

- Calculate the weight/area factor of each kite. (See Project 16.) Does this factor predict the kite's flight characteristics?
- Can you explain in your own words how a box kite flies?
- Can you design a pair of wings so that your box kite will be able to fly in lighter winds?

PROJECT 11: *HEIGHT*

Most kites do not fly straight overhead but at an angle. To observe how efficiently a kite is flying, it is important to make height readings.

Question

How high, on a set line length, does a kite fly?

Materials

Wind gauge (see Project 12)
Kite line of known length. Try 300 ft (91 m).
Tape measure

Procedure 1

1. Sight the kite along the top of the gauge. Place your finger on the thread, turn the gauge and note the angle.

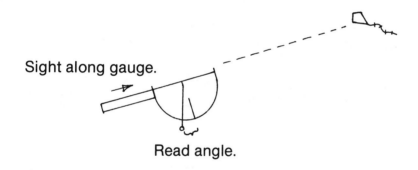

Sight along gauge.

Read angle.

2. Measure the distance from your eye to the ground.
3. Measure the distance from your sighting to a point directly below the kite.
4. Plot these measurements on graph paper and calculate how high the kite is flying.
5. Calculate your own table for different flying angles and lengths of flying line. NOTE: Kite lines have some sag.

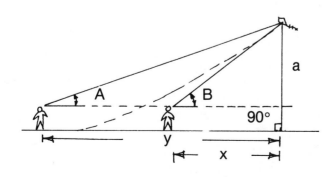

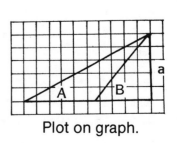

Plot on graph.

Procedure 2

Take the same readings as in Procedure 1 but use trigonometry to calculate the height of the kite.

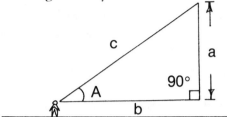

height of kite = a = c x Sin A (length of the kite line known)
height of kite = a = b x Tan A (length of the kite line unknown)

A	Sin	Tan	A	Sin	Tan	A	Sin	Tan
1	0.02	0.02	31	0.52	0.60	61	0.88	1.80
2	0.04	0.04	32	0.53	0.63	62	0.88	1.88
3	0.05	0.05	33	0.55	0.65	63	0.89	1.96
4	0.07	0.07	34	0.56	0.68	64	0.90	2.05
5	0.09	0.09	35	0.57	0.70	65	0.91	2.15
6	0.11	0.11	36	0.59	0.73	66	0.91	2.25
7	0.12	0.12	37	0.62	0.75	67	0.92	2.36
8	0.14	0.14	38	0.62	0.78	68	0.93	2.48
9	0.16	0.16	39	0.63	0.81	69	0.93	2.61
10	0.17	0.18	40	0.64	0.84	70	0.94	2.75
11	0.19	0.19	41	0.66	0.87	71	0.95	2.90
12	0.21	0.21	42	0.67	0.90	72	0.95	3.08
13	0.23	0.23	43	0.68	0.93	73	0.96	3.49
14	0.24	0.25	44	0.70	0.97	74	0.96	3.49
15	0.26	0.27	45	0.71	1.00	75	0.97	3.73
16	0.28	0.29	46	0.72	1.04	76	0.97	4.01
17	0.29	0.31	47	0.73	1.07	77	0.97	4.33
18	0.31	0.33	48	0.74	1.11	78	0.98	4.70
19	0.33	0.34	49	0.76	1.15	79	0.98	5.15
20	0.34	0.36	50	0.77	1.19	80	0.99	5.67
21	0.36	0.38	51	0.77	1.24	81	0.99	6.31
22	0.38	0.40	52	0.78	1.28	82	0.99	7.12
23	0.39	0.42	53	0.80	1.33	83	0.99	8.14
24	0.41	0.45	54	0.81	1.38	84	1.00	9.51
25	0.43	0.47	55	0.82	1.43	85	1.00	11.43
26	0.44	0.49	56	0.83	1.48	86	1.00	14.30
27	0.45	0.51	57	0.84	1.54	87	1.00	19.08
28	0.47	0.53	58	0.85	1.60	88	1.00	28.64
29	0.49	0.55	59	0.86	1.66	89	1.00	57.29
30	0.50	0.57	60	0.87	1.73	90	1.00	–

PROJECT 12: *WIND GAUGE*

Materials

Pattern from page 55
Cardboard
12" x 1" x 1/8" (30.5 cm x 2.5 cm x 3 mm) length of wood (may be cut
from a meter stick)
Glue
16" (41 cm) thread
Needle
Small fishing weight (sinker)
Table tennis ball

Construction

1. Glue wind gauge to cardboard, cut out, and fold.
2. Glue wind gauge to wood. Saw a small slit 1/8" (3 mm) deep at the "0" point.
3. Thread the 16" (41 cm) length of thread through the table tennis ball, using the needle. Tie or glue one end of the thread to the ball. Attach the small weight at other end of the thread and fit the thread into the slit in the gauge. The weighted end of thread should fall slightly below the gauge. This is the plumb line.

Procedure

Take a reading by pointing the gauge into the wind. Holding the device away from your body, read the wind range. Make sure that the plumb line is at "0."

NOTE: The gauge and scale make only approximate measurements.

Think About It

• What elements of your wind gauge contribute to error?
• What elements of your measuring technique can contribute to error?

1

Fold along line.

2

3 Saw slit.

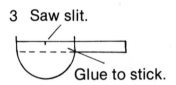

Glue to stick.

4

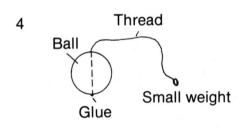

Thread
Ball
Glue
Small weight

5

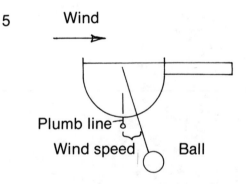

Wind
Plumb line
Wind speed
Ball

PROJECT 13: *WIND VANE*

Materials

Pattern from page 55
Cardboard
1 – 18" x 1/4" (46 cm x 6 mm) dowel
Thumb tack
Compass
Tape and glue

Construction

1. Glue pattern to cardboard. Cut out vane and fit together.
2. Pin vane to dowel. Make sure that vane can spin freely.

Tape

Procedure

Place wind vane in an area that is clear of wind turbulence. It is normal for wind to change directions frequently in some locations.

Observe the wind direction, using your compass and wind vane.

Compass

PROJECT 14: *WIND*

The energy source for a kite flight is the wind. This energy source is in constant change, affected by the local terrain as well by global weather patterns.

Hills, buildings, and trees cause turbulence that makes flying a kite difficult. To eliminate ground turbulence, the standard measurement for wind is made at 33 feet (10 m) above the ground. Wind increases in velocity with higher elevation.

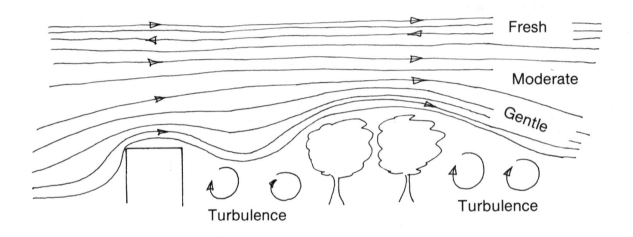

In 1806 Admiral Sir Francis Beaufort developed a rating system for the accurate recording of wind speed. This system was developed for sailors, but has since been modified for use on land.

BEAUFORT NUMBER	M.P.H	WIND SCALE	NAME
0	0-1	Smoke rises vertically	CALM
1	2-3	Smoke drifts slowly	LIGHT AIR
2	4-7	Leaves rustle	LIGHT BREEZE
3	8-12	Small flags fly	GENTLE
4	13-18	Trees toss, dust flies	MODERATE
5	19-24	Small trees sway	FRESH
6	25-31	Large branches sway	STRONG
7	32-38	Trees in motion	MODERATE GALE
8	39-46	Twigs break	FRESH GALE
9	47-54	Branches break	STRONG GALE
10	55-63	Trees snap	WHOLE GALE
11	64-72	Widespread damage	STORM
12	73-82	Extreme damage	HURRICANE

NOTE: 1 MPH equals 0.447 M/sec equals 1.61 Km/h

Question

When is it possible to fly kites, and what affects local wind patterns?

Materials

Wind gauge
Wind vane
Paper and pencils

Procedure

Keep a log on temperature, humidity, and wind conditions, noting any weather patterns. Make individual readings and add weather information from the press.

DAY	TEMPERATURE RANGE	WIND RANGE	WIND DIRECTION	WEATHER NOTES
SUNDAY				
MONDAY				
TUESDAY				
WEDNESDAY				
THURSDAY				
FRIDAY				
SATURDAY				

Think About It

- Are there parts of the terrain (lakes, hills, etc.) in your area that affect the weather in your area?
- Are there times of the year when it is easier to fly kites? What correlations can you make between wind characteristics and the large weather patterns associated with seasonal change?
- Try flying your kite in the upslope winds on the windward side (side toward the wind) of a hill. Try flying your kite on the side of a hill where the wind sweeps down. What differences do you find? What ways can you discover to fly your kite successfully under different conditions?
- Where do you most notice the effects of wind turbulence? How does it make your kite behave?
- What weather conditions are dangerous to aviation?

PROJECT 15: *ASPECT RATIO*

The aspect ratio of a kite, or of any aircraft, is a dimensionless term used to designate the stability of a craft. The aspect ratio relates the width (span) and the length (chord) of the kite. If a kite is very long and narrow, it will have a low aspect ratio and may lift very efficiently but be difficult to stabilize. A nearly square or round kite will have a high aspect ratio: it may be quite stable but difficult to maneuver. As you build different kites in this project, you will notice a continuum of aspect ratios, with unstable kites at the low ratio end, more maneuverable kites toward the middle, and stable but hard to fly kites at the top.

The aspect ratio is expressed as

$$R = \frac{b^2}{A}$$

Where R = aspect ratio
b = span (width)
A = area of the wing or kite

For example, the aspect ratio of a rectangle two meters by three meters:

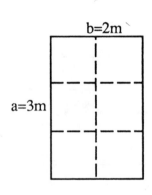

b=2m

a=3m

$$R = \frac{b^2}{ab} \qquad R = \frac{(2m)^2}{(2m \times 3m)} = \frac{4m^2}{6m^2} = .\overline{666}$$

F(newtons) = m (kilograms) x 9.8 meters/second²

or

$$\text{Pressure in pascals} = \frac{\text{Force in newtons}}{\text{Area in square meters}}$$

Question

Is the aspect ratio related to how a kite flies?

Procedure

Compare the flying performance of a number of kites. Calculate the aspect ratio of each of these kites. What correlation do you find?
NOTE: For more information on calculating the area of your kites, see the next project.

Think About It

- Does the aspect ratio help you predict how a kite will fly?
- Should you consider the aspect ratio when designing a kite?
- The aspect ratio expresses only one variable of the many which determine how a kite will perform. What others have you discovered?

PROJECT 16: *WEIGHT TO AREA FACTOR*

One way to evaluate how well a kite will fly is to determine its weight to area factor (m). This rating is found by dividing the kite's weight in kilograms (W) by the area in square meters (A). You can use this factor as you select materials to design your kite to determine if it will be aerodynamically sound.

$$m = W/A$$

Question

Is a weight to area factor a practical rating for a kite?

Procedure

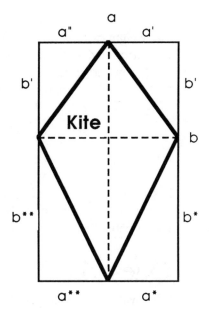

You will need a tape measure, a spring scale, and your kite.

Weigh the kite with the spring scale and divide this reading by the kite's area.

For a diamond-shaped kite, consider that the diamond (your kite) is contained by a rectangle. The area of a rectangle is (a)(b) = Area. Notice that the sides of your kite are diagonals which each bisect rectangular sections of the large rectangle. Measure the sides and angles of your kite, and the sides and angles of the bisected rectangles (a*b*, a'b', a"b", a**b**) to verify this.

IF (a*b*) + (a'b') + (a"b") + (a**b**) = ab, then

1/2 [(a*b*) + (a'b') + (a"b") + (a**b**)] = 1/2 ab

Therefore, the area of your diamond-shaped kite is one half the area of a rectangle which contains it.

If your kite is triangular, the formula for its area is 1/2 base x height:
A = 1/2 bh.
If your kite has an irregular surface area, plot the kite on graph paper to determine the area.
If your kite is a box kite, remember that it has four lifting surfaces.

Determine the weight/area factor (m) for several kites.

Think About It

• Does the weight/area factor give an indication of how a kite will fly?
• How can you use this factor when you design a kite?
• What differences do you notice among kites with different m?

PROJECT 17: *WIND SPEED AND LIFT*

The wind speed (velocity) has to generate a lifting force greater than the weight of the kite for the kite to fly.

The standard textook formula for lift is

$$L = \tfrac{1}{2} L_C \rho S V^2$$

where L   = Lift force
      L_C = Lift coefficient
      ρ = Air density
      S   = Area of kite (square meters)
      V = Wind velocity (m/s)

The weight of the kite (W) is related to its mass (M) by W = Mg, where g is the acceleration due to gravity (g = 9.8 m/s²). Also, typical values for the lift coefficient and air density are 0.6 and 1.3 Kg/m. With these values, the lift formula can be rearranged to predict wind speed for a kite to fly:

$$V = (5 \text{ meters/second}) \sqrt{M/S}$$

For a kite to fly, the wind speed must be equal to, or greater than, the value given by this formula. Note that for this formula to work, the mass must be in kilograms and the area must be in square meters.

Question

What wind speed is required to fly a kite?

Procedure

Use a scale or balance to find the mass (M) of a kite, in kilograms, and divide this number by the area of the kite in square meters (S). Calculate the wind speed required for the kite to fly.

Test the above formula by flying the kite in a wind with the calculated velocity.

You may want to make several small kites of varying weight and area, and fly them in front of a box fan. Use your wind gauge from Project 11 to determine the wind speed at different settings. Fly each kite at different wind speeds. Then calculate the wind speed that the formula would predict is optimum for each kite.

Think About It

- Did the kite fly in the wind speed your calculations predicted it would? If you have access to a rheostat, how great is the range of wind speed that will fly each kite?
- A box fan only approximates the way natural wind would affect your kite outdoors. How does it differ? How could you make this "artificial wind" more representative of natural conditions?
- If you were designing a kite for very light breezes, or for high winds, how would you vary your design?

PROJECT 18: *FORCE*

Newton's third law of motion states that for every action there is an equal and opposite reaction. The wind exerts force on the kite's surface which, because it is tethered, exerts force (tension) on the kite line.

Question

What is the force exerted on a kite at any one wind speed?

Materials

Spring scale
Kite (You may want to use the small kites you made in Project 17. If so, you will also be able to use the box fan from Project 17.)
Wind gauge

Procedure

Measure the pull on the line in kilograms with a spring scale, and multiply this value by "little g," the acceleration due to gravity (9.8 meters/second²). Calculate values for a number of wind speeds using the same kite.

If you are using your setup from Project 17, also calculate values for the same wind speed on different kites.

$$F(newtons) = m \text{ (kilograms)} \times 9.8 \text{ meters/second}^2$$

or

$$\text{Pressure in pascals} = \frac{\text{Force in newtons}}{\text{Area in square meters}}$$

Think About It

- What does the magnitude of the pressures you are measuring tell you about the materials you can use to build kites?
- Does a different kite in the same wind speed give you a different pressure measurement? What does this tell you about kite design for different wind speeds?

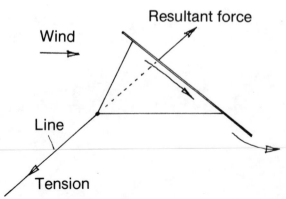

APPENDICES

MATERIALS FOR A KITE

Sail – 24" x 30" (61 cm x 76 cm) plastic trash bag or newspaper

Tail – Plastic trash bag

Tape – Filament or packaging

White glue

Dowels

Flying line – 20# test cotton, polyester, or nylon

Tubing – 3/16" (4 mm) inside diameter (ID)

Framing line – 30# test for some designs

Tools

Protractor

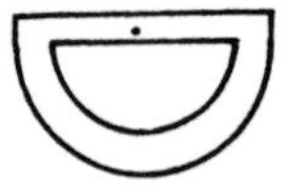

Spring scale

Small saw or hacksaw blade

Tape

Pen or pencil

Meter stick and tape measure

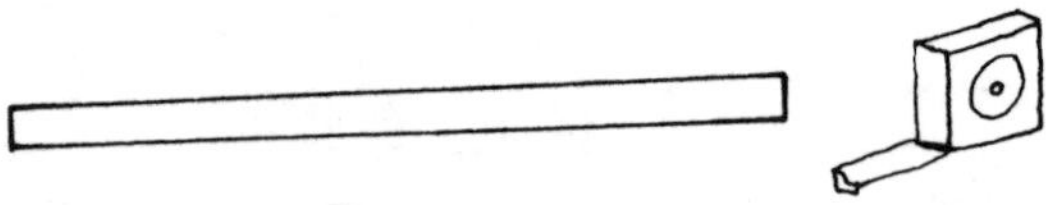

Sharp scissors

Large needle

Leather punch

WHEN TO FLY A KITE

Kite flying has no season, so kites can be flown all year. Fly any time the wind is right and there are no storms. Spring is the traditional kite flying season, but some spring winds are too strong. Remember, if the birds are not flying, do not fly a kite.

Not all kites fly in all wind conditions. Refer to Project 14 for more information about wind.

General Range

MATERIAL	WIND (m.p.h.)	
Light paper	4-12	(Light to Gentle)
Light plastic	8-24	(Gentle to Fresh)
Heavy plastic	13-31	(Moderate to Strong)
Light cloth	8-31	(Gentle to Strong)
Heavy cloth	13-31	(Moderate to Strong)

KITE TYPE	WIND (m.p.h.)	
Sled	6-18	(Light to Moderate)
Diamond	6-18	(Light to Moderate)
Delta	6-18	(Light to Moderate)
Box	13-31	(Moderate to Strong)
Fighter	4-12	(Light to Gentle)

WHERE TO FLY A KITE

Fly kites where they do not create a hazard to you or others. Trees and buildings create turbulence that makes kite flying difficult. The best flying areas are in the open—at the beach or a park.

Wind

Turbulence

HOW TO FLY A KITE

To launch a kite, let out some line, and with your back to the wind, offer the kite upwards. Hold the kite away from you so as not to block the wind. When the wind catches it, slowly pump on the line. Do not pull hard, since this will usually cause the kite to dive. Keep feeding out line until the kite reaches the desired altitude. Do not let the line run through your fingers rapidly or you may receive a burn or cut. Your end of the string should be tied securely to a reel or winder of some type.

If the kite does spin or dive, release tension on the line so that the kite can right itself and land safely. Pulling hard on the line will only increase the dive or spin.

If the kite refuses to fly, check to see if the wind conditions are suitable, whether there is too little or too much tail, or if the bridle loop is in the correct position.

Although running with a kite is regarded by many as part of kite flying, it is not necessary for a good launch. It is hard to see a kite when you are running with it. The kite could easily get hurt and so could you. If you must run, choose smooth terrain.

Sometimes ground level winds are light, while up higher they are stronger. Watch rising smoke, flying kites, or leaves on tall trees for an indication of this situation. Winds near the ground can be turbulent because of local trees or buildings. If the winds near the ground are turbulent or light, it is best to have an assistant help with a "high launch." The assistant should be downwind 15 to 30 meters, holding the kite with the flying line taut. On a given signal, launch the kite with a slight upward push. As the kite rises, pump and release further line.

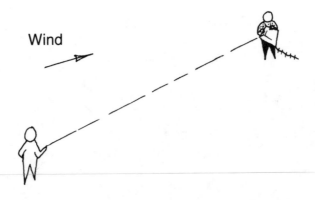

Wind

SAFETY CODE

–NEVER FLY YOUR KITE NEAR POWER LINES. If a kite becomes tangled in power lines, LEAVE IT THERE. Notify your local utility company of the exact location.

–Kites should be flown in flat, open areas away from buildings and roadways. FLY ONLY WHERE THE KITE WILL NOT CREATE A HAZARD.

–Never fly a kite during an approaching storm or in rain, as WET LINE WILL CONDUCT ELECTRICITY.

–DO NOT USE WIRE AS FLYING LINE.

–Always observe local air safety regulations. Avoid flying in air traffic patterns close to airports.

–Large kites can be dangerous and require extra care. Wear gloves when flying large kites.

–Do not let line run through your fingers at a fast rate, as this will result in a burn or cut. Wrap line securely around a large dowel or piece of broomstick. Wear gloves when flying a strong-pulling kite.

–Carry a folding scissors for cutting snarls. Blades and sharp pointed objects can be dangerous.

–Do not throw heavy objects at an entrapped kite. Loosen the line, or cut it, if necessary, so the kite can fly itself free.

–Remember that everyone has the same right to the sky. Keep a safe distance from other fliers.

GLOSSARY

Acceleration: the ratio of the change of **velocity** of an object to the time in which the change occurs. Negative acceleration, or decrease of velocity, is called **deceleration**.

Aircraft: any weight-carrying structure for navigation of the air, designed to be supported either by its own buoyancy or by the dynamic action of the air against its surfaces.

Airfoil: any surface, such as a **wing** or a rudder, designed to obtain reaction upon its surfaces from the air through which it moves.

Angle of attack: the angle formed by the **chord** of the wings with the airfoil's line of flight (also known as the angle of incidence).

Arc: a portion of a curved line, or the path of something that describes such a line.

Aspect ratio: the relationship between the width (**span**) and the length (**chord**) of a kite or other airfoil.

Chord: the line traced from the leading edge to the trailing edge of an airfoil; loosely, its length.

Deceleration: negative acceleration, the loss of **velocity**.

Dihedral angle: the angle between two planes.

Dowel: a round, smooth wooden rod, available in a variety of lengths and widths.

Drag: wind resistance, or the total force on an airfoil due to the air through which it moves.

Equilibrium: a state in which forces are so arranged that they balance.

Force: the effect of **mass** times **acceleration**; a vector whose direction is the same as the direction in which the object accelerates.

Frame: the structure which provides the shape of the aircraft.

Keel: the assembly which projects below the bottom of the craft to provide stability.

Launching angle: the angle the aircraft assumes on taking flight, also known as the angle of entry.

Lift: the force opposed to **gravity**, which is perpendicular to the wind and in the plane of symmetry.

Mass: a measure of the amount of matter.

Pitch: lateral deviation from the line of flight (see drawing, page 11.)

Roll: lengthwise deviation from the line of flight (see drawing, page 11.)

Slack: lacking tension, relaxed, as a slack line.

Span: the maximum lateral distance from tip to tip of an aircraft.

Stability: that condition of a body which causes it to be restored to its original **equilibrium** after a disturbance.

Stall: a situation of flow breakdown (turbulence) in which lift has been lost.

Static angle: the angle at which the kite is held by the bridle, with no wind on the kite.

Taut: the opposite of **slack**; sustaining maximum tension.

Tetrahedral: having four sides, each side an equilateral triangle. Tetrahedral kites were developed by Alexander Graham Bell, and can be made of one or many tetrahedrons.

Turbulence: the flow of a fluid (such as air) in which the velocity at a given point changes rapidly in magnitude and direction.

Velocity: the ratio of the displacement of an object to the time interval required for the displacement.

Wing: a portion of an aircraft's surface which functions as an airfoil.

Yaw: deviation from the line of flight by angular motion about the horizontal or vertical axis of an aircraft (see drawing, page 11.)

RESOURCE LIST

Gayla Industries
P.O. Box 920800
Houston, TX 77292-0800

Gayla inflatable kite
Kites

American Kite Fliers Association
6636 Kirkley Ave.
McLean, VA 22101

Kiting journal

United States Plastic
1390 Neubrecht Road
Lima, OH 45801-3191

Plastic tubing

Edmund Scientific
101 E. Gloucester Pike
Barrington, NJ 08007

General science supplies

Kitemaking Supplies (contact for price list)

Kitty Hawk Kites, P.O.Box 340, Nags Head, NC 27959

The Kite Port, 2112 Crowfoot Dr., Lafayette, IN 47905

The Unique Place, 525 S. Washington, Royal Oak, MI 48061

Kite Kompany, 33 W. Orange, Chagrin Falls, OH 44022

Kite Kraft, School Haus Sq., Frankenmuth, MI 48734

Kite Site, 3101 M St., Washington, DC 20007

High Fly Kite Co., 30 West End Ave., Haddonfield, NJ 08033·

Into The Wind, 1729 Spruce St., Boulder, CO 80302

Great Winds Kites, 402 Occidental Ave. S., Seattle, WA 98134

Windplay, 232 S.W. Ankeny Alley, Portland, OR 97204

NOTE: Dowels and plastic tubing are available at most hardware stores, hobby shops, or lumber yards.

BIBLIOGRAPHY

Anderson, John Jr. FLIGHT, McGraw-Hill, New York 1985.

Brummitt, Whyatt, KITES, Golden Press, New York 1971

Bushell, Helen, MAKE MINE FLY, AKA, Melbourne, Australia 1977.

Greger, Margaret, KITES FOR EVERYONE, Self-published, Richland, WA 1984.

Halliday, D. and Resnick, R. FUNDAMENTALS OF PHYSICS, Wiley, New York 1970.

Hosking, Wayne, KITES: AUSSIE STYLE, Self-published, Houston 1982.

Hosking, Wayne, KITEWORKS 1, Self-published, Houston 1986.

Ito, Toshio and Komura, Hirotsugu, KITES, Japan Publications, Tokyo 1979.

Markowski, Michael, UTRALIGHT FLIGHT, Utralight Publications, Hummelstown, PA 1982.

Moulton, Ron, KITES, Pelham Books, London 1978.

Pelham, David, KITES, Penguin, New York 1976.

MEASUREMENT CONVERSIONS

Length and Distance

Inches (in)	1 in = 25.4 mm = 2.54 cm
Feet (ft)	1 ft = 30.5 cm
Yards (yd)	1 yd = 0.914 m
Miles	1 mile = 1.61 km
Millimeter (mm)	1 mm = 0.0394 in
Centimeter (cm)	1 cm = 0.394 in
Meters (m)	1 m = 3.28 ft
Kilometers (km)	1 km = 0.62 miles

Speed

Miles per hour (m.p.h.)
$$1 \text{ m.p.h.} = 1.61 \text{ km/h}$$
$$= 1.47 \text{ ft/sec}$$
$$= 0.447 \text{ m/sec}$$

Kilometers per hour (km/h)   1 km/h = 0.621 m.p.h.

Weight and Mass

Ounces (oz)	1 oz = 28.3 g
Pound (lb)	1 lb = 454 g
Gram (g)	1 g = 0.035 oz
Kilogram (kg)	1 kg = 2.2 lb
Slug	1 slug = 32.2 lb

Force

Newton (nt)	1 nt = 0.225 lb

Physical Properties

Acceleration due to gravity	9.81 m/sec^2	= 32.2 ft/sec^2
Air Density	1.29 kg/m^3	= 0.002378 slug/ft^3

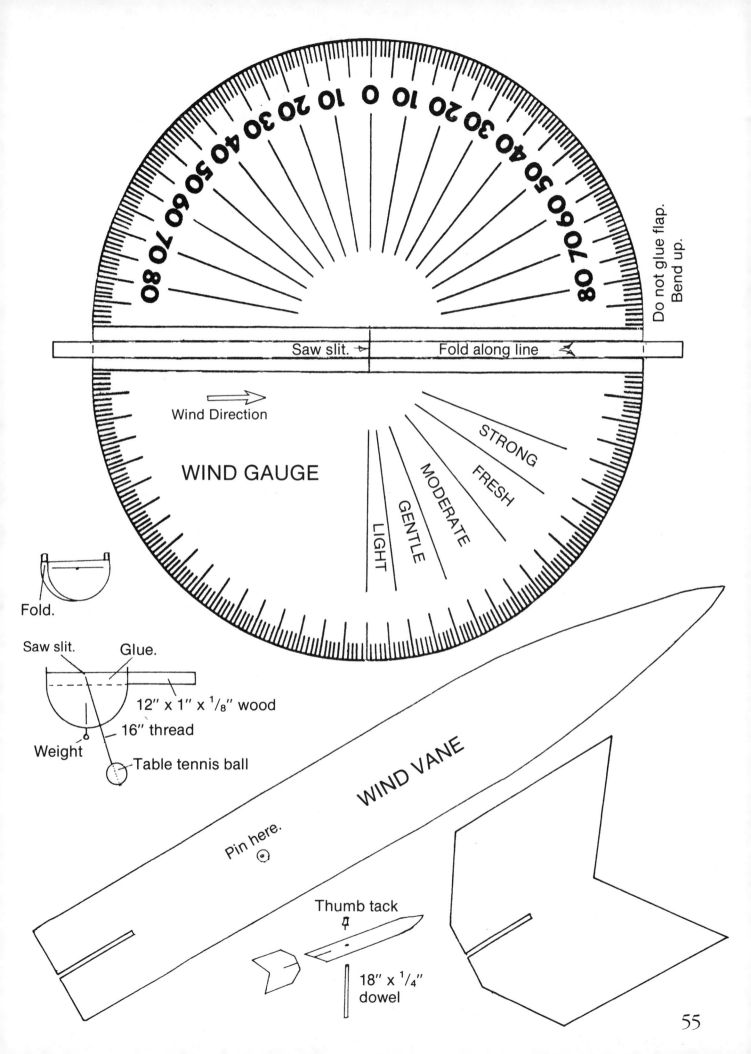

WIND GAUGE

Wind Direction

STRONG
FRESH
MODERATE
GENTLE
LIGHT

Saw slit. Fold along line

Do not glue flap. Bend up.

80 70 60 50 40 30 20 10 0 10 20 30 40 50 60 70 80

Fold.

Saw slit. Glue.

12" x 1" x 1/8" wood

16" thread

Weight

Table tennis ball

WIND VANE

Pin here.

Thumb tack

18" x 1/4" dowel

55